MW01671986

A Visitor's Guide to the Historic Quincy Mine

An Historical tour of the Quincy Mining Company site, the Nordberg Hoist, and the Quincy and Torch Lake Railroad.

LAWRENCE J. MOLLOY

ISBN-13: 978-0-9791772-0-0

Published by Great Lakes Geoscience, LLC, Ontonagon, Michigan

Key to Photographic Credits

The following abbreviations are used to give credit to individuals and organizations who provided photographs for this book:

HAER - Photograph courtesy of the Library of Congress, Prints and Photographs Division, Historic American Engineering Record. These photographs are followed by a reproduction number that identifies the individual photograph in the collection, such as *MICH,31-HANC,1-37.* Many photographs have an individual donor noted, such as:

LGK/HAER -Louis G. Koepel donor

QMC/HAER - Quincy Mining Company donor

HAER drawings and drawings based on HAER drawings are identified by the creator of the drawing and the date and number of the drawing in the collection.

LJM - Photograph from the collection of Lawrence J. Molloy

MTU/CCHC - Photograph courtesy of the Michigan Technological University Archives and Copper Country Historical Collections.

SAM - Photograph courtesy of the State Archives of Michigan

Other individual contributors are noted in detail on their photographs.

Cover Photo Credits

The line drawings on the cover were created from photographs contributed by:

Quincy Number 2 Shaft-Rockhouse, circa 1890, *SAM*

Quincy Number 2 today, LJM.

Author's Note

History is dynamic and not static, it changes as our understanding of events changes. In compiling this book I've used data from many different and sometimes conflicting sources. In some cases I've discovered that the same item may have two entirely different names, and both may actually be correct. In these cases I have placed the alternate information in parenthesis () near the term. I have made every effort to balance the facts and present as accurate a portrayal of the history of the Quincy Mining Company and the Quincy and Torch Lake Railroad as possible in the limited space of this book. I apologize in advance for any missing or apparently inaccurate information.

Lawrence J. Molloy
May, 2007

A Visitor's Guide to the Historic Quincy Mine

Table of Contents

Preface

The drive from Detroit to the Keweenaw is a long one. I know that I'm almost there when, south of Chassell, I can look out in the distance and see Quincy No. 2 standing like a beacon guiding me to the Keweenaw. I was not born a native of the Keweenaw, but my research and love for it's history has allowed me to become an adopted child. I've been lucky enough to have led historical tours in the Keweenaw and discuss its history with numerous classes from elementary school to colleges. My interest led to the publication of my first book, Copper Country Road Trips, an historical tour from Copper Harbor to the Porcupine Mountains. Part of that book provides a brief tour of the Quincy mine site. I was asked to take those pages and expand on them to give a better understanding of the history that surrounds the Quincy Mine site. This book is not intended to be a comprehensive analysis of the Quincy, there are several outstanding volumes listed in the reference section that do that. Rather this book was designed to supplement the surface and underground tours offered by the Quincy Mine Hoist Association. The photographs and text should help put a face on those who made the Quincy famous. There is more to the Keweenaw than the Quincy. But for many of us the Quincy has a special place in our hearts. I hope this book helps to explain why.

Acknowledgments

Copper Country Road Trips had only been available for a few weeks when it was suggested that I write a follow-up book that would focus on the Quincy Mining Company. A quick check with my wife Nancy and family assured me that they were once again willing to lose me for weeks to my books and my computer. Without their support and patience this work might never have been completed. I'd like to thank those researchers who went beyond the call to help me find a favorite photo or obscure fact. Eric Nordberg and the staff of the Copper Country Historical Collections at Michigan Technological University have provided hours of aid. The staff of the State Archives of Michigan were patient enough to pull file after file for me. Thanks to reviewers Glenda Bierman, Jim Brooks, Dr. Theodore J. Bornhorst, Deborah Molloy, Dave Piotrowski, Judy Matteson, Charles Pomazal, Dr. George Robinson, and Ed Yarbrough who not only read the original draft for me but supplied me with comments that were in themselves an education for me. A special thanks to Jim Hird and John McKana for their help in identifying the equipment on display at the Quincy site. Finally I'd like to thank Ed Yarbrough and all of the staff of the Quincy Mine Hoist Association for their help and patience in allowing me to roam up and down, over and literally under the ground to get a better understanding of the Quincy Mining Company and the Quincy and Torch Lake Railroad.

A Visitor's Guide To The Historic Quincy Mine - Tour Map

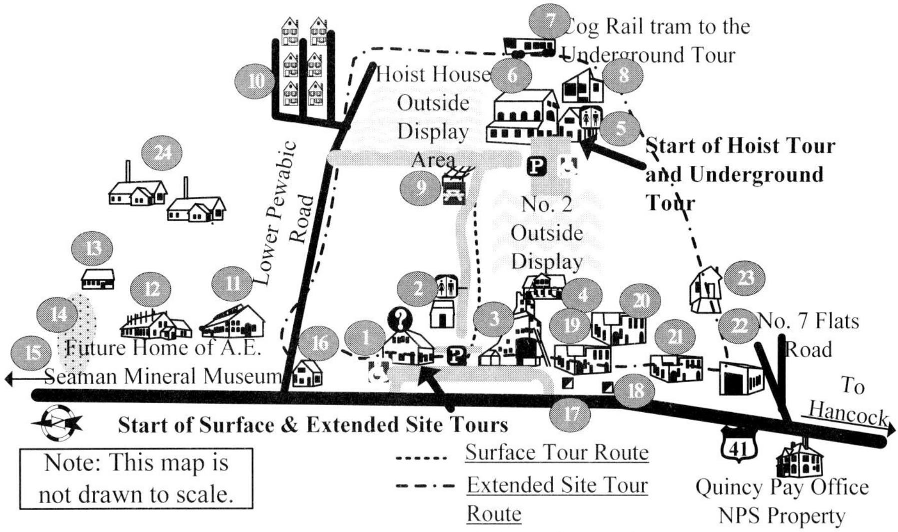

The locations listed below in bold are part of the Quincy Mine surface tour. Others may be visited free. Some sites are on private property. Use caution when walking around ruins and other sites of the Quincy Mine. Numerous hazards exist. Do not enter any abandoned buildings or ruins. Watch out for trip hazards, open holes, lose rocks, and unstable ruins.

1 Supply Office (1893) (Gift Shop and Ticket Office)
2 Oil /Powder House (1893) (Restrooms)
3 **No. 2 Shaft-Rock House House (1908)**
4 Old No. 2 Hoist House (1882-1894)
5 **Old No. 2 Hoist House Museum (1894-1920)**
6 **No. 2 Hoist House - Nordberg Mine Hoist Museum (1918-1931)**
7 **Cog Rail Tram (constructed in 1996)**
8 No. 5 Boiler House (1912)
9 Cooling ponds for the No. 2 hoist steam engine (1918)
10 Remains of Company Community of Lower Pewabic
11 Machine Shop (1900)
12 Blacksmith Shop (1900)
13 Ruins of North Quincy (No. 6) Dry House (pre-1890)
14 Remains of North Quincy (No. 6) Shaft (1891-1931)
15 No. 8 Shaft, Head Frame and Hoist House (worked 1896-1931 and 1976-77)
16 Mine Captains Office (ca. 1860)
17 "Man Engine" Shaft (1866-1892) (Bat entrance)
18 No. 4 Shaft (Capped) (1856-1909)
19 Ruins of No. 4 Boiler House (1882)
20 Ruins of No. 4 Hoist House (1885)
21 Ruins of No. 7 Boiler House (1898)
22 Locomotive Engine House or Roundhouse, Q&TLRR (1889)
23 Locomotive Water Tower, Q&TLRR (1898)
24 Ruins/smokestacks of No. 6 Boilers

One of the earliest photographs of the Quincy - This 1875 photo shows, from left to right, the No 1, 2, 3, and 4 shaft houses. These structures basically just covered the shaft. Two hoist houses, between No1 and 2 and between No 3 and 4 are also visible. *(QMC/HAER,MICH,31-HANC,1-8)*

Perhaps the most recognizable feature in the Keweenaw today is the No. 2 Shaft–Rockhouse of the Quincy Mining Company standing atop Quincy Hill overlooking Houghton and Hancock. From 1846 until 1945 the Quincy Mining Company was one of the major copper producers in the Keweenaw and, at times, one of the major producers in the world. It not only mined the copper ore here, but built mills to crush it, a smelter to purify it, and a railroad. The Quincy and Torch Lake, to carry ore from the mine to the mill. The **Quincy Mine Hoist Association** (Q.M.H.A.) offers excellent tours of the No. 2 Shaft-Rock House and the 1918 hoist house with its Nordberg hoist engine. It also offers a short tram ride that leads to a tour of the underground workings of the Quincy Mine. This tour book supplements those tours and provides a broad historical view of the mine site as well as a more in-depth look at specific features. The book also offers visitors the option of a longer extended site tour that involves walking much of the mine location.

Early History

Archeologists believe that native people were mining copper in the Keweenaw as long ago as 5,000 B.C. The first European mining effort to mine copper in the Keweenaw took place near Ontonagon in 1771. However, the real effort to mine copper here began as a result of a land swap between two new states. When Michigan was admitted to the Union in 1837 it received the Upper Peninsula in place of a strip of land that it really wanted around Toledo, now Ohio. In 1841 Michigan asked its first state geologist, Douglas Houghton, to survey and report on this then little-known territory. Houghton's report mentioned the vast iron and copper holdings he had seen and soon Michigan's copper rush had begun. Years before the gold rush in California prospectors headed to the Keweenaw hoping to pick up their fortune from the copper lying on the ground. They soon learned, however, that it would take companies and capital to take the wealth that lie in the ground. Companies took out leases on land and then sent miners to explore it. Unfortunately the leases sometimes overlapped. The Portage Mining Company and the Northwestern Mining Company of Flint both held leases for the same land. In order to resolve the dispute they met in 1846 and formed the Quincy Mining Company. As the company grew, it acquired the lands of many adjoining mining companies, beginning with the Pewabic Mining Company (formed in 1853, acquired in 1891) just north of them. Quincy also acquired some of the lands of the St.

Under Construction—The 'Modern' No.2 shaft-rockhouse as it was built around the still working older No. shaft-rockhouse. Note the men on the steel rigging high above the 1882 hoist house. *(HAER MICH,31-HANC,1-15)*

The No. 4 shaft, 1895 - This building is taller than the earlier shaft houses but still performed only one task, hoisting. It performed no rock processing. To its left is the 1881 machine shop, and the 1894 No2 shaft-rockhouse.*(QMC/HAER,MICH,31-HANC,1-29)*

Mary's Canal Mineral Land Company (which obtained land in exchange for developing the locks at Sault Saint Marie), the Mesnard Mining Company (formed in 1859, acquired in 1897); the Pontiac Mining Company (formed in 1860, acquired 1895), and the Franklin Mining Company (formed in 1857, acquired in 1908) until it owned the entire Pewabic lode.

Early mining efforts at the Quincy were not commercially successful. It wasn't until the Pewabic lode, a layer of copper rich rock, was discovered by the Pewabic Mining Company on land that it later found belonged to Quincy in 1856 that it looked like the Quincy mine might be profitable. Even then the first dividend wasn't paid to stockholders until 1862. Mining at the Quincy in the 1850'-60's was primitive. Miners climbed ladders into the ground, worked by candlelight and at first used hand-cranked windlasses to raise kibbles or large buckets of ore to the surface. Large pieces of **native copper** (nearly pure) called **mass copper** were often chiseled into smaller pieces before lifting. At the surface smaller pieces of mass were placed into barrels for shipment (**barrel copper**). The remaining ore was sent to a rockhouse to be separated from the rock. Here cam-driven Cornish stamps would crush the ore and free additional copper. Copper was loaded onto boats and shipped to Pittsburgh, Cleveland, and Detroit to be smelted. What you see today is truly a different world than that faced by the early Quincy Mine.

The Quincy was the major copper producer in the United States during the 1860s and was second only to the Calumet and Hecla Company for the remainder of the 1800s. Quincy paid dividends so often that it earned the nickname 'Old Reliable'. It mined the Pewabic Lode until 1945 when the mine closed its underground works for the last time. By then its shafts were over 9,000 feet deep. From 1945 to 1967 the Reclamation Plant at Mason allowed the company to produce copper from waste stamp sands that had been dumped into Torch Lake decades earlier. Following a brief exploration project in 1976-77 in conjunction with the Homestake Copper Company (a subsidiary of the Homestake Mining Company) the Quincy Mining Company ceased mining copper.

The Surface Tour

Begin your tour at the Q.M.H.A. Gift Shop, which was originally the **Quincy Supply Office**. The sandstone-building was built in 1893 at a cost of $18,000. Behind it is the 1893 **Oil House**, now the rest rooms for the location. Walk

Construction Complete—The No. 2 Shaft-Rockhouse as it looked after construction was completed in 1908. It combined hoisting and processing by making the building taller and aligning the processing with the shaft. The small stone building is the 1882 Hoist House for the original No. 2 shaft. *(HAER MICH,31-HANC,1-17)*

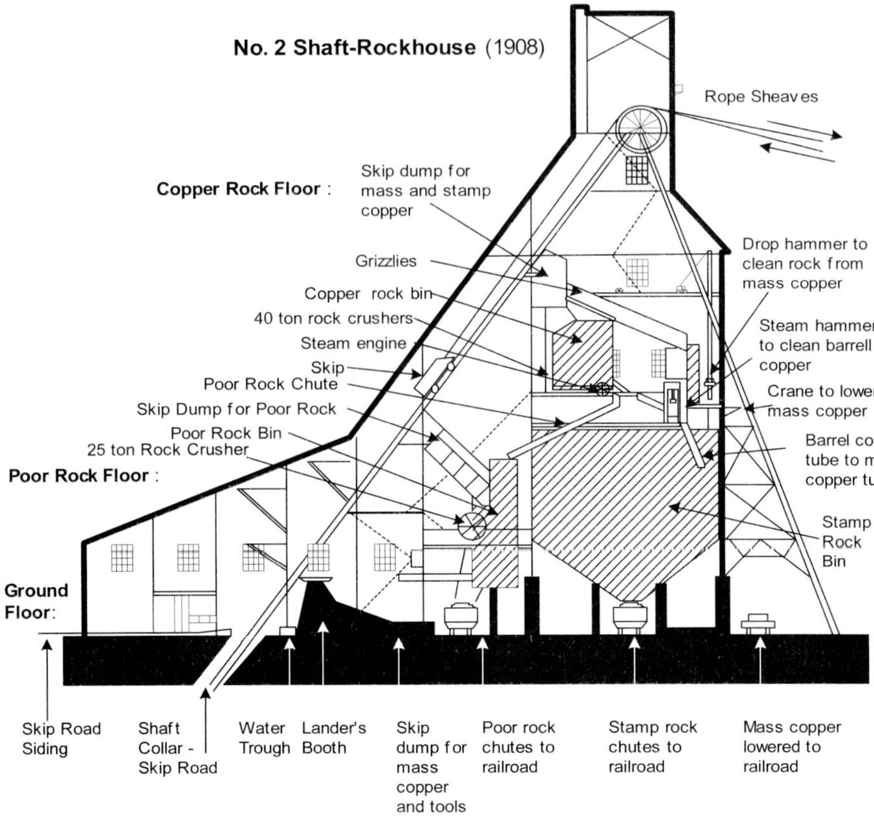

This drawing is based on HAER drawing, MI-2,19/34, Durward W. Potter, Jr. , 1978 and Richard K. Anderson, Jr. , 1979.*(LJM)*

across the driveway to the **No. 2 Shaft-Rock House**. Inside the building is a continuously playing videotape that explains the details of the operation of this structure. Built in 1908 by the American Bridge Company, this headframe replaced a many gabled wooden structure built in 1895. The No. 2 shaft was opened in 1858 and closed in 1931. Three of the hoist houses, whose hoists or engines wound the wire rope that raised and lowered 'skips' full of ore, still remain. The small building directly behind the No. 2 headframe is the 1882-1895 Hoist House. In a straight line some distance behind are the hoist houses from 1895 and 1920, which will be discussed later.

The No.2 Shaft Rockhouse : Taller is better

Simply put, a shaft house covers a hole or shaft in the ground. It controlled the lowering and raising of men and ore from the mines' underground workings. Early shaft houses were wooden houses built over a shaft. In the 1860's wheelbarrows were used to move the ore from the shaft house to kiln houses.

Here a wooden bed 24 feet square and 4 feet high was covered with 4 to 6 feet of copper rich ore. The wood was set fire and the heat would crack the rock. This process, called calcining, made it easier to remove the copper from the rock. Water was sometimes poured on the ore to help it crack. Mechanical crushers were introduced later. In 1873 a tram road was built from the shaft houses to a single rockhouse to process the ore before it went to the stamp mill.

In 1892, at the No. 6 shaft, Quincy first introduced the idea that the ore processing (crushing and sorting once done at the rockhouse) and ore hoisting in skips (done in a shaft house) should be in the same building called a shaft-rockhouse. In the early shaft-rockhouses ore crushing and sorting took place in a wing of the building set to the side of the shaft house. In 1908 it was decided that this new process could be even more efficient if the shaft-rockhouse were made taller so that gravity could assist in moving ore from place to place rather than having to use machines to move material side wards. The No. 2 shaft-

The No. 6 shaft-rockhouse - Shown here in 1920, No. 6 was constructed in 1891-92. It was the first Quincy structure to combine the hoisting function (in the tall, center section) with ore processing that was performed in a wing to the side. *(LGK/HAER,MICH,31-HANC,1-37)*

Steam hammer cleaning barrel copper, circa 1895-1900 - The steam hammer inside of No.2, similar to this one in No.6, smashed ore to help remove the copper from the rock. *(QMC/HAER, MICH, 31-HANC, 1-34)*

rockhouse provides an excellent example of how the redesign worked. The current structure is the last structure built on the No. 2 shaft. The base of the building is rises 147' from the ground. The building dips at an angle to follow the angle of the lode. It angles 20 degrees for a horizontal distance of about 60 feet and then 54 degrees for an additional 50 feet. Behind it are two of the original eight pulley stands that supported the wire rope as it ran from the hoist house to the shaft-rock house.

Quincy was always trying new plans to see if efficiency could be increased. These trials worked in many cases. In 1904, for example, 28 men processed 900 tons of rock per day at the No. 7 shaft. By 1907 it took only 5 men to process over 1800 tons in a taller, reorganized structure.

Inside a shaft-rockhouse

Inside the shaft-rockhouse is the shaft collar that surrounds the opening to the mine. Two sets of parallel railroad tracks, called skip roads, carried skips (cars) into and out of the mine. The skips were hoisted by means of 1 5/8" wire rope that was pulled over head sheaves at the top of the shaft-rockhouse, then supported by a series of pulley stands as the rope was then joined to the large hoisting engine located to the rear of the shaft-rockhouse. In the No. 2 there were three distinct levels of activity. On the ground floor stands the landers' booth. Here the lander, the person in charge of the skip once it entered the shaft-rockhouse, could communicate by a series of bells with the filler in the mine below and the hoist operator hundreds of feet away in the hoist house. The lander could operate a lever that would alter the spacing on the rails inside the shaft-rockhouse and dump the skip at the desired level. Skips with mass copper (large pieces of copper free of rock) or heavy tools to be unloaded would be stopped at the ground level. Also, water skips could be dumped into the water trough here. There are also a series of cranes here to assist in changing from rock skips to man cars to raise and lower workers into the mine.

Skips that contained rock that had no copper content, called poor rock, were dumped into a bin on the poor rock floor. Here this rock, and the poor rock sorted from the copper floor above, would flow into a poor rock bin. Some of the rock from this bin would be sent to the 25 ton rock crusher and turned into loose aggregate that could be used in roads, rail beds, etc. The rest of the poor rock would be sent by rail to poor rock piles similar to those that still dot the Keweenaw today.

Going to work, Quincy No. 2 shaft, circa 1910s - Note the young boys in the man car as well as the miners' sunshine and carbide lamps. *(MTU/CCHC Wilbert Maki donor)*

Skips that contained copper bearing ore and smaller pieces of mass copper would be dumped at the copper rock floor. As the ore was dumped from the skip it first passed over a series of grizzlies or bars approximately 20" apart. Ore that then passed through another set of bars set 2 3/4" apart was ready to be sent to the stamp mill. It fell directly into the large stamp rock bin (ore bin). Ore larger than 2 3/4"

Grizzlies, circa 1900 - This photo from one of the Quincy shaft-rock-houses, shows ore resting on the grizzlies (bars) that help to sort the material by size *(MTU/CCHC, Neg. 01216)*

but smaller than the 20" spacing fell into a bin that fed the 40 ton rock crushers. From there it fell into the stamp rock bin. Some ore with heavy copper in it had to be sent to a steam powered hammer to clean the rock from the copper rather than to the crushers. Copper from the steam hammer was sent down a chute that held barrel and mass copper until it was loaded onto a rail car. Ore larger than 20" was sent to a drop hammer. Here a steam driven winch raised a weight or hammer that fell by gravity onto the ore and broke the copper free from the rock. Smaller pieces of this copper were stored in the barrel and mass copper bin while larger pieces of mass copper were lowered outside of the shaft-rockhouse by a crane to the railroad below.

Tour Note: *Those following the shorter surface tour may wish to explore the displays near No. 2 as you walk from No 2. shaft-rockhouse to the Nordberg Hoist and the underground tour center. Those items are discussed next in this tour book. If you are walking the longer site tour route you may wish to go to page 40 now and examine the following pages as you approach the Nordberg hoist from a different path. Be sure to examine the displays near No 2. before leaving the Quincy location.*

Quincy Mine Equipment Displays: No. 2 Outside Display Area

The Quincy Mine site contains two outdoor display areas of mining equipment that was used in the Quincy Mine as well as equipment from other mining sources, including the Calumet and Hecla Company's Kingston and Osceola mines in Keweenaw county, the White Pine mine near Ontonagon, and the Cleveland Cliffs Iron Mining Company (CCI) and mines on the Marquette iron range. The No. 2 Outside Display Area lies between the Number 2 Shaft-Rockhouse and the Nordberg Hoist building. See page 1 for the general location and page 11 for a detailed layout. The Hoist House Outside Display Area is east of the Nordberg Hoist House (see page 26 for details.) Some of the items on display near No. 2 include:

Concentrate Transfer Bucket: This was used in a mill to move concentrate or processed material high in copper content.

Quincy Mine - No. 2 Outside Display Area

This map is not drawn to scale.

See page 1 for a complete site map.

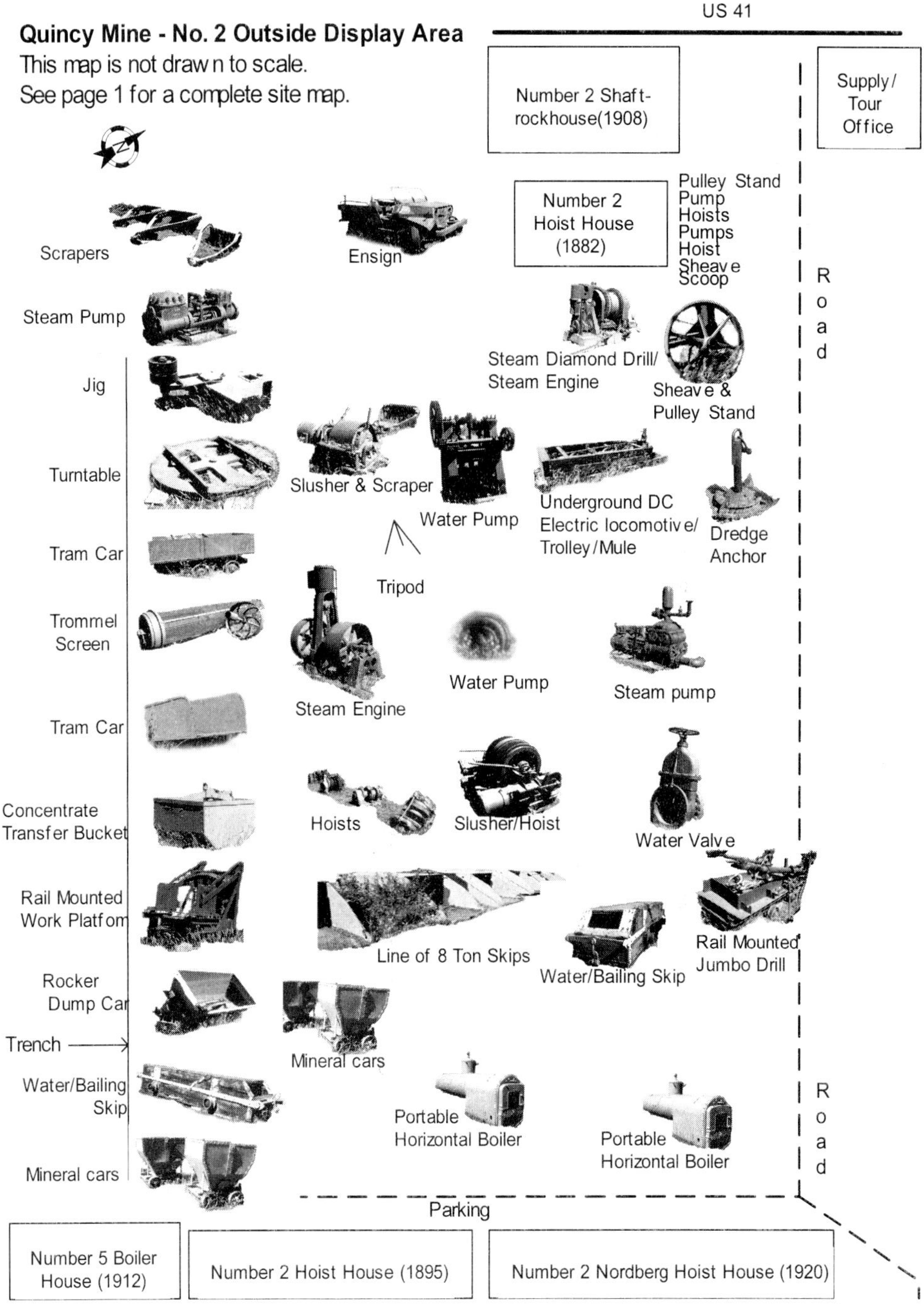

Loading a Skip - Here the rock from the tram cars had been dumped into a chute or pocket cut above the shaft. The loader or filler opens the chute and fills the skip in the Quincy in the 1920s. Note the size of the wire rope and the bolt attached to the front of the skip. *(QMC/HAER, MTU, John F. Campbell collection, MICH,31-HANC,1-122)*

Dredge Anchor: This anchor was used by the dredges that reclaimed copper from Torch Lake.

Ensign: The Ensign 100 is a Jeep type vehicle first produced by the Copper Range Company in its maintenance shops in Painesdale, Michigan in 1971. It was used in the White Pine Mine in Ontonagon County. The success of this, the only locally produced vehicle in the Keweenaw, led to the development of the Coratec division of the Copper Range Company which marketed the Ensign.

Hoists: Hoists with a single drum are 'tuggers" while double drum hoists are called 'slushers'. Hoists such as these would pull wire ropes attached to large metal blades, called "scrapers" that would slide or pull rock and material into chutes, cars, etc.

Jig: Jigs were used in a mill to concentrate ore by placing the mineral on the screen and then rocking the screen under water or pulsing water through the screen to help remove impurities.

Locomotive, Motor, Trolley, or Mule: Known by different names at different times and in different areas, it is a direct current (DC) underground locomotive powered by a pulley rolling along an overhead trolley wire. It would pull three to five tram cars full of ore to the shaft where the ore would be hoisted in a skip. Quincy began using locomotives in late 1901. Quincy began experimenting with battery powered locomotives in 1917.

Mineral Cars: These cars were used in the smelter to move material to charge it into the smelting furnace.

Portable Horizontal Boilers: These devices, from CCI, were used to produce steam on the surface for the wide range of steam powered devices used at the mine site.

Rail Mounted Jumbo Drill: This car supported several air drills and allowed multiple holes to be drilled at once.

Rail Mounted Work Platform: By moving up and down this platform raised workers in the drift to assist in placing timber, etc.

Rocker Dump Car: These are often found in iron country. Cars like this are still used today.

Scrapers: These large blades were pulled with slusher hoists to move ore down a stope or to move ore horizontally in an adit. They sometimes also helped load the rock into tram cars by pulling ore.

Sheave and Pulley Stand: At one time there were eight(8) large towers or pulley stands between the Number 2 Hoist House and the shaft-rockhouse. Large metal pulleys (wheels or pulley sheaves) supported the wire rope as it traveled to and from the hoist to the shaft-rockhouse. At the top of the shaft-rockhouse are two head sheaves, one for each skip road, that supported the rope as it lowered and raised the skip in the mine shaft.

Diamond Drilling - The steam engine and winch in the North Display Area, from CCI, would have been used as part of a diamond drilling rig like the one shown here, circa 1890s. These drills took samples of the rock for exploration. *(MTU/CCHC, Roy Drier Collection, Neg. 00468)*

Skips: Skips were used to hoist the ore and poor rock from the mine to the surface. In 1863 the Quincy began replacing kibbles with skips. (Kibbles were early, large , barrel shaped metal buckets that were used to hoist ore. Some mines still use buckets called hoisting buckets for

Slusher and Scraper - This photo, circa 1920s, shows a double drum slusher or hoist attached to a scraper to move ore. The scraper would be pulled to the miner and then away from him by pulling the wire rope first from one drum and then the other drum of the slusher *(LJM)*

specific tasks.) These small skips had a capacity of 1 1/2 to 2 tons. Skip capacity was increased as power to hoist them increased so that the 6 ton capacity skip (approximately 9 to 10' long) was introduced in 1894, the 8 ton in 1902 (approximately 12 1/2' long), and the 10 ton skip (approximately 15' long) in 1920. The skips in this row could each hoist approximately 8 tons of rock.

Slusher & Scraper: One section of wire rope connected one drum of the slusher to the front of the scraper blade and another section ran from the other drum of the slusher through a sheave block behind the blade and then back to scrapper blade. By pulling on the appropriate end of rope the slusher was able to pull the scraper and rock towards the slusher and then return it to the pile to drag more rock.

Steam Engine: Many machines, including diamond drilling machines, were operated by steam and different types of steam engines were used by mining companies.

Steam Diamond Drill: This equipment powered a diamond drilling operation. Mining companies would use diamond drills to explore new ground. Drilling could be done vertically above ground with a device similar to this one, or done horizontally with a drill in a drift or stope of an existing mine.

Steam Pump: As steam enters and leaves one side of the pump, the other side of the device would pump water by sucking water into a cylinder and then forcing it out.

Tram Car: Small cars such as these were often pushed manually by workers in the mine. Different types of cars could haul rock, equipment, etc. Tram cars came in various shapes, sizes, and designs. Some were balanced and hinged to allow for easy tipping of the car.

Tripod: A simple stand used to support a rock drill.

Trommel Screen: The screen in this cylinder was used in a mill to help sort material by size.

Turntable: In some drifts (corridors) tram cars tracks ran directly into the shaft opening. In other drifts the tram car had to be spun on a turntable such as this

prior to dumping.
Water/Bailing Skip: Quincy began using skips like these in 1903 to remove water from the mine. With a capacity of 1,300 to 1,600 gallons, these skips were lowered into sumps at the bottom of the shafts where water had settled or had been pumped. These skips hoisted the water to the surface. Water bailing normally took place on Sundays so as not to interfere with production.
Water Pump: This is a slurry pump used in a mill.
Water Valve: Valves such as this controlled the flow of water.

As you complete your tour of the No. 2 Display area you are approaching the 1894 Hoist house. It is now an exhibit area for the Q.M.H.A., as well as, the meeting point for the underground tour. Attached on its left is the 1918 hoist House and the Nordberg hoist, the world's largest steam hoist.

Tram cars - Cars such as the one on display were used to move a variety of objects in cars of various shapes. Note the large piece mass of copper above the car and the mine captain dressed in white, c.1920s. *(MTU/CCHC, donor The Copper Range Company, Neg. 01730)*

A view from the No. 2 shaft-rock house taken in the 1920s showing, from the right, the No. 5 boiler house, the 1895 hoist house, the 1918 hoist house, and the box-shaped cooling ponds at the far left. The insert shows a similar photo taken in August, 2000. Note how the homes have disappeared and how trees have reappeared on Quincy Hill.*(MTU/CCHC, Nils Eileetson, donor, Neg. 03361, Insert-LJM)*

The Hoist House Tour—The Nordberg Hoist

The year was 1917. The Number 2 shaft had reached a depth of almost 1 1/2 miles on the incline. The hoist, an engine used to raise and lower items into the shaft, was housed in the "new" hoist house, the **1895 No. 2 Hoist House**. The smaller of the two current buildings, the complex cost $90,000. It replaced the older hoist and hoist house that was built in 1882 and is located just behind the present Number 2 shaft-rockhouse. When the E. P. Allis company built the hoist in 1894 it was the largest one they had ever built. It's twin steam cylinders rotated a hoisting drum, 26 feet in diameter that could hold up to 7,500 feet of 1 1/2" wire rope. It could move two 6-ton skips, in balance at a rate of 2,500 to 3,000 feet per minute. But now an even more powerful hoist was needed. The Quincy Mining Company had paid dividends for 48 continuous years and needed to go deeper and hoist larger cars faster if it was to continue its record of production. The company was also concerned about the cost of the coal that was used to heat the boilers to make the steam that drove the many hoists and machines at the mine site. In order to address these needs, and also to show Quincy's faith in the future of the mine, a new, more efficient hoist, and hoist house were planned that would be the largest steam powered hoist in the world.

The 1895 E.P. Allis Hoist and House— This photo from c. 1900 shows the earlier Allis hoist that was used at No.2. It was installed in the 1895 hoist building which is now the Quincy Hoist Tour building and museum. The timbered ceiling aids in determining where the hoist once stood. *(HAER MICH,31-HANC,1-52)*

1918 No. 2 Hoist House

It was hoped that the hoist would be available in 1917, but World War I and its limits on non-war related production delayed the delivery of the hoist until 1919. Work on the hoist house was begun by the Maclean Construction Company of Chicago in 1918. Following the plans of J. M. Hoff, a civil engineer also from Chicago, the building was completed in 1919. The selection of an outside firm to build the house and its design, were major changes for Quincy. Quincy had always designed and built their own buildings. The early hoist houses were wooden structures. As time progressed they were made of poor rock that was removed from the mine. Later buildings, such as the 1895 No. 2 hoist house, were made from the red sandstone found at local quarries near Jacobsville and the Portage Entry. But this new building was to be a jewel in the Keweenaw. To support the weight of the 1,765,000 pound hoist, they created the largest block of reinforced concrete ever poured at that time. The 54' by 84' block contained 3,200 cubic yards of material, including 8 tons of steel

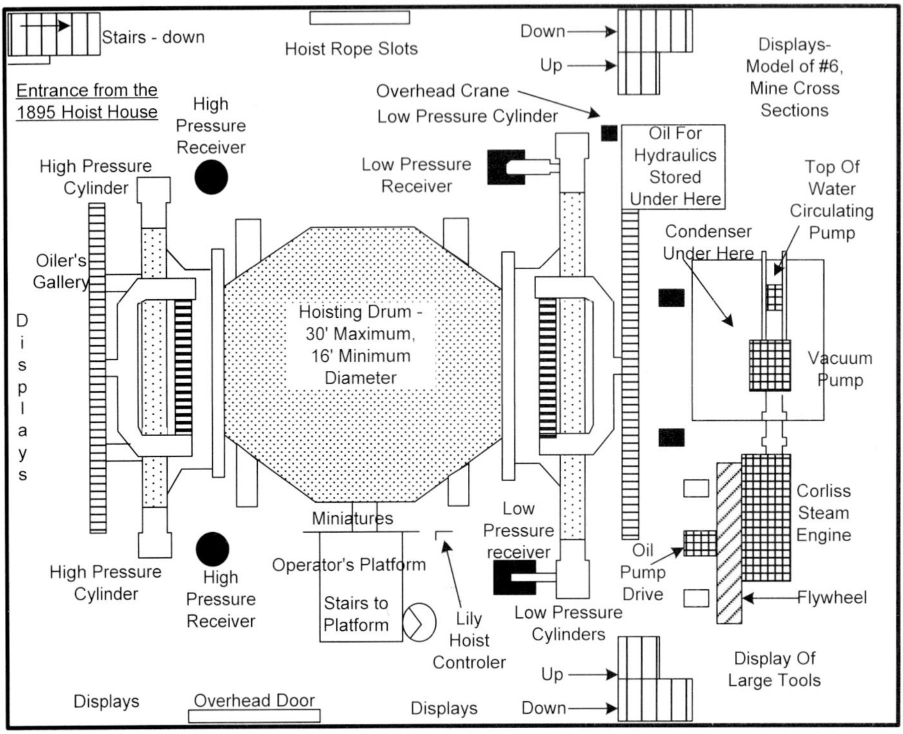

This drawing is based on HAER drawing, MI-2,14/34, Durward W. Potter, Jr. , 1978. *(LJM)*

rails and 8,000 feet of 1 1/2" steel hoisting rope. The foundation alone cost $42,700. The part of the building that houses the hoist itself is 72 feet by 76 feet and is 82 feet high. Attached is a small building to house the condensing equipment for the hoist. In the front of the hoist house are two large slits (one smile and one frown) where the hoisting rope would leave the building and travel on pulley stands to the Number 2 shaft-rockhouse. In the rear is an overhead door through which every piece of the hoist had to fit as it was assembled in the house. The entire structure, including the roof, is built of reinforced concrete and is fireproof. This eliminated the fire problem that had destroyed several earlier buildings. A decorative red brick veneer was placed on the outside walls of the building. The windows were made of ribbed glass, thicker than normal glass. The white tiles that were placed on the inside walls were imported from Italy. Some of them may still be seen today. The roof was originally covered by green glazed, Imperial Spanish ceramic tile from Mexico. While this roof was later removed several pieces of the tile are on display in the hoist house. These finishing touches were unheard of when Quincy built the house. The entire building, not counting the foundation, cost $57, 900.

The Hoist

Quincy ordered their new hoist from the Nordberg Manufacturing Company of Milwaukee in 1917. It took a year to assemble and became operational on November 9, 1920. At that time the No. 2 shaft had reached an incline depth of 7,750'. The assembly process was itself an impressive effort requiring oversized tools, some of which are on display, and the installation of an overhead crane that is still in place today. The crane alone cost $6,500 and was only used to build and maintain the hoist. The Nordberg engine cost $181,600. The total cost of the engine, auxiliary equipment, the hoist house and the equipment in it was $371,000. The Number 5 boiler plant, built in 1912, was selected to provide the steam for the hoist. It stands today just south of the 1895 hoist house. In order to keep down-time to a minimum during the switch from the 1895 to the 1918 hoist, the new hoist house was built immediately to the north of the old, just slightly off-line with the shaft.

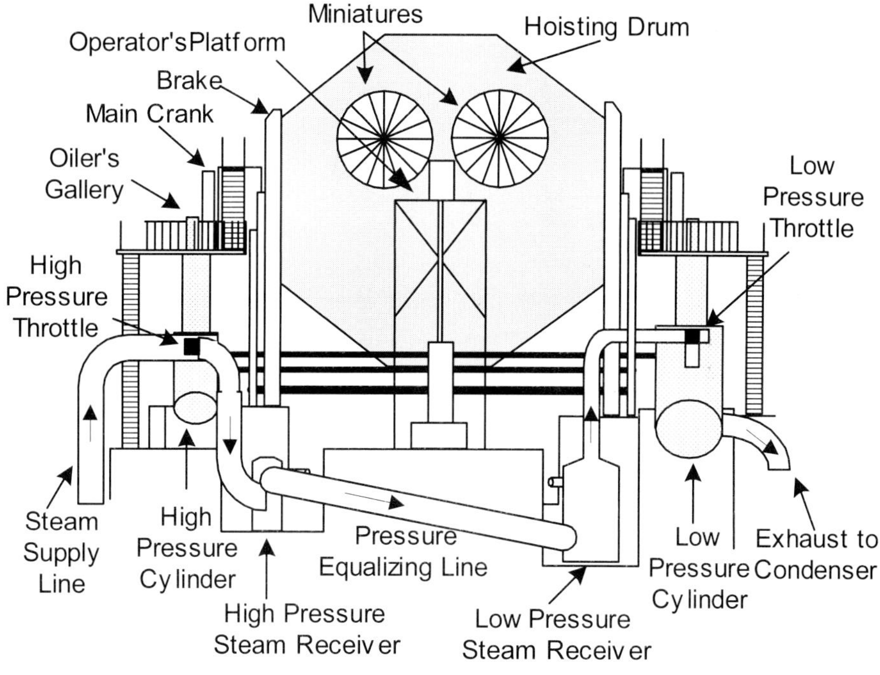

This diagram illustrates the major features of the hoist and traces the flow of steam through the hoist. It is based on HAER drawing, MI-2,13/34, Jon R. Carter, 1978. *(LJM)*

Early Hoist Photograph *(LJM)*

How the hoist worked

The Nordberg Hoist is actually four steam cylinders, four steam engines working as one engine. A steam engine functions by using alternating intake and exhaust valves to allow steam to enter a chamber, expand inside of the chamber, and use the force of the expansion to push a piston as the steam expands. The double-acting pistons used here can be pushed both up and down in the cylinder by the expanding steam. The entire engine takes 60 feet by 54 feet of floor space and is 60 feet tall. It is a cross-compound steam engine, an engine where steam is used twice. The earlier Allis hoist was a duplex hoist and was not compound. Steam was used only once and then released. This common technique accounts for the "choo-choo-choo-choo" sound one hears near steam powered trains. Here is how the Nordberg Hoist worked:

1. The boilers in the Number 5 boiler plant boiled water to create steam.
2. The steam entered the hoist via the steam supply line at a pressure of 160 psi (pounds per square inch).
3. High pressure throttles (valves) controlled the flow of steam into the intake bonnet for the two high pressure cylinders (located on the south side of the hoist nearest the smaller 1895 hoist house). In each of these cylinders was a piston with a 32" diameter/bore that moved a stroke or distance of 66".
4. Two intake (inlet) valves opened and closed to allow steam to enter the cylinder, one on each side of the piston. On the other side of the cylinder are

two exhaust valves that allowed steam that had expanded to leave the cylinder. These valves opened and closed in a series of distinct stages to move the piston and therefore move the drum. The first stage is the admission stage. During admission one of the steam intake valves, such as the front valve shown here as a white circle, opened to allow steam while the other three remain closed.

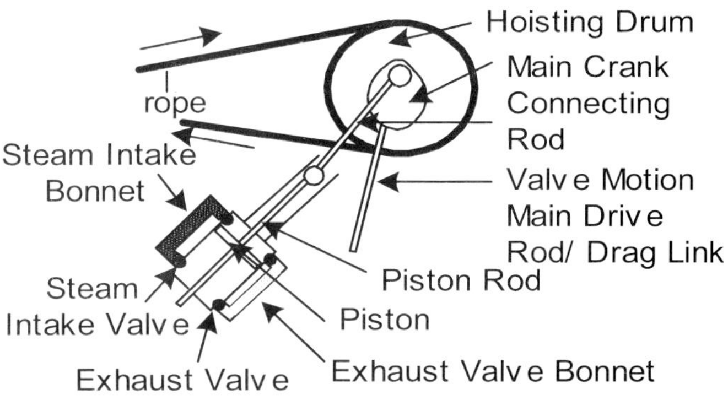

This illustration shows the major parts of one of the cylinders of the hoist used in turning the drum. This drawing, not to scale, is based on HAER drawing, MI-2,15/34, Richard K. Anderson, Jr., 1979.*(LJM)*

5. The second stage is the expansion stage. The steam intake valve was closed. Now, as soon as the exhaust valve began to open, the steam began to expand inside of the cylinder. As it expanded it pushed against the piston and moved it, which in turn moved the piston rod.

6. The piston rod is attached to a connecting rod. The connecting rod is also attached to a main crank that is itself fastened to crank shaft through the center of the drum. As the piston and the piston rod moved in parallel up and down, the end of the connecting rod swung in a circle and rotated the main crank which turned the drum. As the drum turned, one segment of rope was wound over the top of the drum while another segment of rope came off of the bottom of the drum allowing one skip to be raised while another was lowered at the same time, keeping the skips or cars in balance. A large metal rod, called the valve motion drive rod, or the drag link, is also attached to the crank. It alternated the opening and closing of the valves.

7. The third stage is exhaust. While the piston was still in motion the opposite exhaust valve, the rear valve shown here as a white circle, opened and began to allow steam that was ahead of the piston to leave the cylinder.

8. Steam left each high pressure cylinder via one of two exhaust valves and the exhaust bonnet. It was collected in one of two high pressure receivers below the floor. A line connecting each high pressure receiver equalized the pressure in them.

9. The fourth stage is compression. The steam from the high pressure cylinder

Auxiliary Steam Engine—as seen today (*LJM*)

enters the cylinder at 35 psi. This creates a vacuum on the low pressure side of the piston, causing the piston to slow down as it prepares to reverse direction.

10. Steam was now admitted from the opposite end of the cylinder. As the piston reached the end of its stroke the other steam input valve, shown here as a white circle, opened and admitted steam. This steam now pushed on the other side of the piston and the main crank continued to turn. The four stage process continues in the reverse order. If the hoist operator wanted to reverse the hoist, the operator used a lever which, by use of a series of shafts and gears, caused the valves to reverse the cycle. The valve that was open closed and the closed valve opened, thus pushing the piston in the opposite direction.

11. The exhaust from the high pressure side traveled to one of the two low pressure steam receivers. These receivers were also connected by a line to equalize pressure.

12. Steam flowed into one of the two low pressure cylinders located near the north side of the building at 35 psi. Since the pistons in these cylinders must move the same 66" stroke but at a reduced pressure, the cylinder has a wider bore of 60" to help match the effort of the high pressure cylinders. It is this reuse of the steam that makes the hoist a cross-compound steam engine. Valves again allow the steam to move the piston.

13. The steam from each cylinder flowed into a condenser where contact with cold water on the steam created a partial vacuum. This increased the efficiency of the steam in the low pressure cylinder.

14. From the condenser the water is pumped to the cooling ponds near the front of the hoist building. From the cooling ponds the cold water is pumped back to the condenser to be reused.

The Drum

The shape of the drum, thicker in the middle than on the ends, may seem unusual to some visitors. Its cylindro-conical shape creates a drum with 76 turns on its center section and 40.95 turns on the cone on each side. It can carry 10,000 feet of 1 5/8" wire rope, weighing 41,500 pounds, (about 4 pounds per foot) when wound in only one layer over the center cylinder and one of the conical shaped ends. The drum itself weighs 516,000 pounds and is made of 36 sections bolted and trussed together. The rope on the drum can reach 9,620 feet when it follows the inclined angle of the shaft. When the second cone of the drum is used, the drum can reach 13,300 feet on the incline. The four cylinders provide eight impulses of power to the drum per revolution. The drum normally rotated at 34 revolutions per minute.

Operating the Hoist

The hoist operator climbed 33 steps to the operator's platform and spent his entire shift on the platform, controlling the ascent and descent of the skips as well as monitoring the steam, water, and vacuum pressure gauges. The operator watched two large circles called miniatures (hoist indicators) to track the location of the skips in the shaft. One tracks the skip going down and the other the skip coming up. The numbers on the circle indicated the number of the level (horizontal working area) where the skip was currently located. A series of bell signals were used to communicate between the hoist operator and the lander (the person who worked in the shaft-rockhouse whose job it was to control the skips after they entered the building) and those in the mine below who loaded the rock into the skips. Originally there were no telephones to provide verbal communication between the operator and those who needed the skip. The operator had to monitor the miniatures carefully. The only way to raise and lower the skips was to rotate the drum on the hoist. If the brake was applied too slowly it would leave the skip between levels and unable to be loaded until the drum was once again rotated to just the correct position. Since time was money the operator had to be perfect in his use of the controls to place the skips at just the right spot. There was one major incident during 1912 in which the operator of the earlier Allis

Operating the hoist - Hoist engineer Frank B. Nancarrow operating the No. 2 Nordberg hoist in 1925. *(LGK/HAER,MICH,31-HANC,1-57)*

The Lily Controller —as it looks today (*LJM*)

Hoist neglected to stop the hoist. The empty skip began to rush down the shaft. As the loaded skip raced upward, its cable snapped. The sudden release of weight on one side of the pulley caused the tops of four stands to snap off. The skip then plunged downward. Luckily no one was hit during either skip's downward ride. The accident cost Quincy $10,000 in damages and a week to repair the damage. Later, in 1923, the Quincy installed a device, the Lily hoist controller at the cost of $2,500 on the Nordberg hoist. It is located near the stairway to the operators platform. It regulates both the maximum speed of the skip and prevents a skip from going to the bottom of the shaft or the top of the shaft-rockhouse. The controller would reduce the flow of steam to the pistons thus slowing the drum, and apply the brake.

Pumps that serve the hoist

On the north side of the hoist building is a steam engine and several pumps that were used to keep the hoist operating. An oil pump was needed to control the braking system of the hoist. Parallel motion brakes were applied to the drum shaft and were activated by lowering the oil pressure. The brake shoes were large pieces of bass wood. Due to its soft nature, bass wood does not glaze as other woods do. A fixed weight applied the brake, and oil pressure relieved the brake. If the oil pressure failed the brakes would be applied automatically. A water pump and a vacuum pump were used as part of the condenser that took the steam from the low pressure cylinders and condensed it before sending it to the cooling ponds. These ponds were 50 feet by 150 feet and held 340,000 gallons of water. A set of 45 nozzles sprayed 1,800 gallons of water into the air. As the water left the first tank it then entered the second tank for further cooling. The Corliss steam engine on the north side of the building powered these pumps.

Did The Hoist Fulfill Its Goals?

The new hoist could move larger, 10-ton ore capacity skips (total weight 13 tons) at a rate of 3,200 feet per minute, or about 36 miles per hour. Not only did this meet the goal of larger and faster loads, but in its first year it burned 2,400 fewer tons of coal, saving $16,000 in fuel. Unfortunately the hoist was not able to compensate for other increased costs of production, and 1920 was the last year in the string of 53 consecutive years that Quincy paid dividends. The Nordberg hoist ran for 11 years. It hoisted skips full of copper ore, man cars that carried miners and supplies underground, and hoisted water bailing skips to empty sumps at the bottom of the shaft. For 11 years, 24 hours a day, the Nordberg hoist served the Quincy well. When Quincy closed the No. 2 shaft in 1931 the shaft had reached a depth of over 9,000 feet on the incline. Today, the hoist remains an engineering marvel and is still the world's largest steam mine hoist. Thousands of visitors come each year to see it. The 1894 hoist house now has several functions. It houses the offices of the Quincy Mine Hoist Association. It contains many interesting exhibits. One of the favorite displays is the train layout that represents the activities of the Quincy and Torch Lake Railroad.

Quincy Water Cooling Ponds , circa 1920s (LJM)

Quincy Mine Equipment Displays: Hoist House Outside Display Area

Next to the 1918 hoist house is another set of outdoor displays. See page 1 for the general location and page 11 for a detailed layout. The first display area was explained on page 10. The display area near the Nordberg Hoist contains many items, including:

Calyx Drill Cores: These large cores, over 5' in diameter, were cut by a Calyx, a rotary steel drill as part of a mining operation in the iron country of Michigan. They were once on display on the M.T.U. campus. Drill cores used to explore the structure of underground rock are normally 1" in diameter.

Concrete Mixer: Mines used concrete and mortar in many projects.

Compressor: Mines used many compressors, such as this small one made by Curtis Manufacturing.

Flat Car: Cars similar to this were used to move materials on the surface at the mine site.

G. E. Electric Motor: As time progressed Quincy switched many of its devices from steam to electric power.

Locomotive, Trolley, or Mule: This is a Direct Current (DC) electric powered underground locomotive. It would pull three to five tram cars full of rock to the shaft where it would be hoisted to the surface in a skip. (See the No. 2 Dis-

Rock crushers - Machines such as these shown in the famous C&H Red Jacket Shaft-rockhouse, circa 1920's, crushed the larger pieces of rock hoisted from the mine. *(MTU/CCHC, C&H Collection, Box 25, folder 4)*

Quincy Mine - Hoist House Outdoor Display Area
This map is not drawn to scale.
See page 1 for a complete site map.

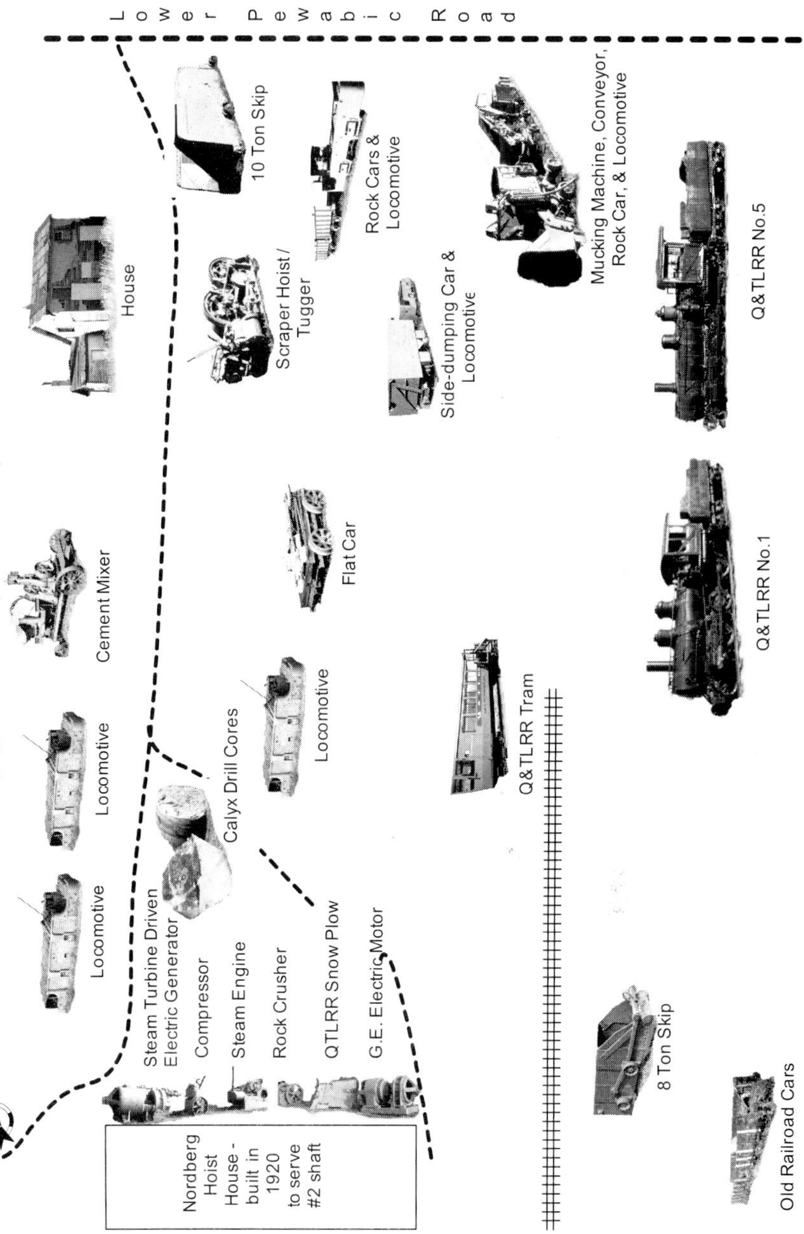

Tramming - No matter the mine, tramming was a back-breaking job, as these trammers in the Tamarack mine show, circa 1920s. *(MTU/CCHC, C&H collection,MS-003, box 25, folder 4)*

play text on page 10 for additional details)
Miner's House: The Quincy Mine Hoist Association is in the process of rehabilitating this home, which is typical of the homes in which many miners lived.
Mucking machine, conveyor, tram car, and locomotive: This combination from the C&H Kingston mine illustrates the later stages of drift mucking. Powered by compressed air, the mucking machine (scoop loader), by Gardner-Denver, would lift the muck or the broken rock onto the conveyor, by Eimco, which would dump it into the rock car. This eliminated the heavy manual labor normally associated with mucking.
Old Railroad cars: These cars once hauled ore and coal for the Quincy and Torch Lake Railroad (Q&TLRR).
Q&TLRR No. 1: The Quincy and Torch Lake Railroad, Engine Number 1. It was named the Thomas F. Mason after the president of the Quincy Mining Company.
Q&TLRR No. 5: Engine Number 5, was built in 1891 for the Hancock and Calumet Railroad. The Q&TLRR acquired it in 1908.
Q&TLRR Snow Plow: This plow was used for snow removal.
Q&TLRR Tram – Opened in 1997, this cogwheel tram travels 2,300 feet from here to the Quincy Mine adit, 360 feet below. Tourists travel from 5 mph on

steep grades, those with a maximum of 37 degrees, to 10 mph on more level grades. Up to 28 individuals ride the tram not only to take the mine tour but to also get a spectacular view of Houghton and the Portage Lake lift bridge.

Tram cars & locomotive: These were used in the Kingston mine to move ore along the level below the stope where the miners were mining the ore to the shaft where it would be loaded into a skip and hoisted to the surface.

Rock Crusher: These were often found in a shaft-rockhouse. Rock would be crushed as it passed down through the opposing heavy metal jaws that were arranged in an inverted pyramid.

Scraper Hoist or Slusher: Hoists such as this one made by the American Derrick Co., would pull wire ropes attached to large metal blades or "scrapers" that would pull ore down the stope to chutes where it was loaded into tram cars.

Side-dumping car and locomotive: Designed and patented by Quincy in 1909, the side-dumping car eliminated the need to stop and manually dump tram cars into the skip at the shaft. The "follower" wheel on one side of the car rode up a small ramp and the hinged car dumped to the side. Cars like this Gandby style side dump car were used in the Kingston mine, and are still used today.

Skip: Skips were used to hoist the ore from the mine to the surface. In the South Display area is an 8 ton and a 10 ton capacity skip. The skips here are from the C&H Osceola mine. (See the No. 2 Display Area on page 10 for in-

Electric Haulage - This 1920 photo shows two young workers near an electric locomotive with a tram car dumping onto a grizzley. *(QMC/HAER,MICH,31-HANC,1-120)*

formation about skips.)
<u>Steam Engine</u>: Steam engines, such as this one by Jewell Manufacturing powered many devices.
<u>Steam Turbine Driven Electric Generator</u>: Steam turned fan blades in the turbine which powered the generator that produced electricity.

Underground tour

Quincy Geology

The key to understanding how and why copper was mined in the Keweenaw lies in its unique geology. About 1,100 million years ago the area was covered by a thick succession of basalt lava flows similar to those in Hawaii today. Hundreds of flows filled the Keweenaw rift to a thickness of about 15 miles beneath present day Lake Superior. As these lava flows cooled the gases in them rose to the top of the flow and created small, almond-shaped holes called vesicles. Vesicles filled with minerals including native copper, are called amygdules. The tops of some lava flows are broken up during eruption. Originally the top of the lava flow was porous and less dense than the massive bottom of the flow. Gravel and sand was deposited between some eruptions. Some time after the Quincy lava flows were deposited, they were compressed by forces with their origin hundreds of miles away. This compression caused the lava flows to tilt towards present day Lake Superior and created faults and fractures. Hydrothermal fluids (hot water plus elements like copper) moved up from deep beneath the surface following the porous tops of the lava flows, inter bedded conglomerate, and fractures. As these hydrothermal fluids rose upwards they cooled and mixed with cool shallow water. This caused **native copper** and other minerals to precipitate from the fluids and fill the open spaces in the rocks at Quincy. The vesicles, open spaces in broken flow tops, fractures

The Quincy is a series of layers of lava flows and interbedded sedimentary rock, all tilted towards Lake Superior. Copper is found in open spaces in the top of lava flows. The Quincy Mine followed the top of a lava flow to extract copper rich rock.

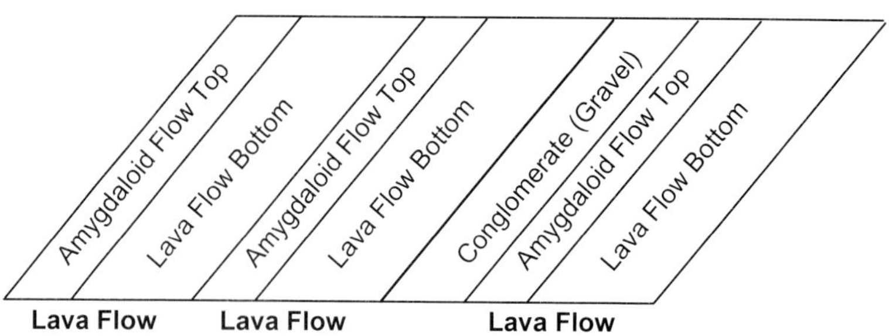

Sample drawing showing layering of material important to mining. *(Based on information from Bornhorst and Lankton, 2006.)*

Quincy miner, c. 1920 - Note the rising angle as the stope is cut upward from the drift. *(MTU/CCHC, U.S. Bureau of Mines, Neg. 35469)*

and faults were filled with many different mineral species, including native copper. A visit to the A. E. Seaman Mineral Museum at Michigan Technological University will allow you to see the many different and beautiful minerals deposited by these hot waters about 1,050 million years ago.

Much later, between 2 million and 10,000 years ago, glaciers covered the area. The glaciers ground up the surface and plucked up native copper exposed at the surface. The malleable native copper was smoothed and flattened into regular shapes called "float copper". Surface weathering produces a green mineral coating on the float copper. You can see a large piece of native copper near US 41 and the gift shop.

As early as 7,000 years ago native people mined the copper in the area. Almost all mining companies, including the Quincy, began their explorations by excavating ancient mining pits. These lodes all dip to the northwest. The angle of descent varies by the lode. The Quincy mine follows the Pewabic Amygdaloid lode as it dips about 50 degrees towards Lake Superior. The lode was about 12 feet thick near the surface and narrowed to three to five feet as the mine reached greater depths. Like many mines the early Quincy hoped to make its fortune by mining large masses of copper. However, it soon became apparent that its real wealth lie in mining the Pewabic Amygdaloid (top of lava flow)

Mass Copper - While the Quincy made most of it's money from the copper found in the small bubbles of the Pewabic Amygdaloid, it also found mass copper deposits such as this one photographed in 1890. Few were as large as this one. Note the candle for light and the pipe for compressed air for drilling. *(QMC/HAER,MICH,31-HANC,1-125)*

that began at the surface of its property and continued over a mile underground.

The Tram

In 1997 the QMHA opened one of the three cog-wheeled railroads in the United States. Trams, as explained on pages 42 and 43, were used to move ore downhill to the mills.

The Quincy Tram, showing the cogs on the track and the operator's console. *(LJM)*

The original Quincy tram railroad carried ore 2,200 feet and declined 500 feet from the top of Quincy Hill down to the original mill on Portage Lake. The modern tram carries up to 28 passengers while it travels 2,300 feet and declines 360 feet to the mine adit entrance. The maximum grade on the tram is 35 percent. Visitors sit facing forward as they go down to the mine entrance and later return to the hoist house, which provides them a magnificent view of the Portage lift bridge and Houghton. The engines of the tram power a series of gears that grip the cogs or teeth on a steel track that sits between the two outside tracks that support the tram wheels, as seen in the photograph. Visitors spend approximately 7 minutes in the tram as it travels at a even pace going down to the mine site at speeds from 5 MPH on the steeper grades to 10 MPH on the flatter grades. Visitors often take scenic photographs from the 35' long, glass enclosed tram as the operator controls it with a joystick type device.

The Adit

An adit is a horizontal opening into a mine. The East Adit of the Quincy, travels some 2,200 feet into the mine. It was first opened in 1894 as a drainage tunnel. Because it was not used for mining, the size of the opening was kept small. From 1976 to 1992 the Department of Mining Engineering at Michigan Technological University used it as a teaching and research facility. Michigan Tech widened the adit to its present size. In 1993 the Quincy Mine Hoist Asso-

Quincy 'Man Engine' circa 1890 - Note the candles on the miner's hats for illumination the ladder, supporting timbers, and the low head room *(QMC/ HAER,MICH,31-HANC,1-113)*

Ready to Blast - Compare this 1920 pattern of blasting holes here with the patterns on display on the mine tour.*(QMC/HAER,MICH,31-*

ciation began using the adit for tours. The mine maintains a fairly constant temperature of 44 degrees at this level.

As you journey into the adit you are passing by about 20 individual lava flows and one sedimentary (gravel) layer. The tour takes you to the area of the No. 5 shaft. You are now approximately 700' feet below the surface. You have 'crossed' under US 41 and are now below the western side of Quincy Hill. The No. 5 shaft or opening, starts at the surface and descends at an angle of about 54 degrees to match the inclination of the layers of copper rich rock. The No. 5 shaft is now capped and sealed at the surface. Shafts were driven into the denser part of the lava flows to provide support and lessen the need for timbers for support. You are at the 7th level of the shaft. A level might be described as a horizontal tunnel below the ground. Early levels were 60' apart but later, as more mechanized forms of mining came into use, the levels grew to as much as 160'.

Opening the Mine

Most Keweenaw mines were begun by sinking the shaft at an angle to follow the angle of the copper lode underground. Later development continues by creating tunnels or drifts. Drifts were driven through ore. In the early days when the Quincy used manual labor and hand drills the drifts were kept small, about 6' high by 7' wide. The drifts near the adit were widened to 8' by 16' to allow for machines. Near the No. 5 shaft is an opening that leads up to the next level. This passage from level to level is called a raise (for moving up) or a winze (for down). They were also used for exploring stopes as well as letting air flow from level to level.

Early Mining Techniques

In the beginning the only way to drive (create) a drift was to use sledge hammers and 18"-24" long drills made of steel to create holes in the rock. Later the holes were filled with black powder. If one man held the drill steel while hitting it with an 4-pound sledge hammer it was called "Single Jacking." A more efficient process involved one man holding the drill while another hit it with an 8-pound hammer. This activity was called "Double Jacking." Usually Double Jacking used two men who alternated blows while a third held the drill steel. "Jacking" referred to the use of skilled miners, most of whom came from Cornwall, England. They were called "Cousin Jacks" since they always seemed to have a cousin named Jack, back home who happened to be an expert miner and needed work.

Q&TRR Engine House and the 1887 rock house, circa 1897. Note the railroad turntable in front of the roundhouse. (*SAM*)

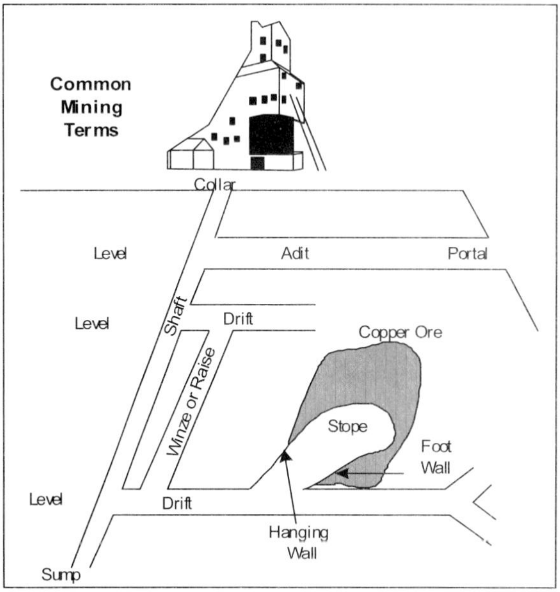

Mining Terms - These terms are used in describing the Quincy. The drawing is not to scale. *(LJM)*

The only light available was that of a candle held onto their soft hat by a lump of clay, or possibly held in an iron candle-holder called a miner's toothpick (Tommy-stick, candle spike). The drill holder rotated the drill between blows to make a round hole in the rock. Once the series of holes were drilled blasting powder would be placed in the holes. When the powder was detonated it broke the rock free. In early days when black powder was

Quincy Tram Road and Rock Houses circa 1890 -Note the dual tracks and cables running past the rock houses and the blacksmith shop far left of the photograph. (*MTU/CCHC, neg. 01618*)

used, blasting was often done on a Saturday afternoon so that the dust could settle on Sunday before work resumed on Monday.

As the drifts spread out from the shaft the miners would discover copper in several forms. Most of the copper at Quincy was discovered in the top of the volcanic flows as amygdaloid copper. As they moved from volcanic layer to layer they might also cross cracks, or fissures, where large masses of copper were found. While the Quincy made most if it's profit from amygdaloid copper it found many masses of copper, some weighing several tons. Miners would start from the drift and mine upwards following the angle of the mineralized rock towards the next level. The mining areas are called stopes. You can see a stope near the No. 5 shaft. There are stopes along the top of the Pewabic flow for 2 miles to the northeast and for some 7,000 feet below your feet on a incline that follows the top of the lava flow. As the stope grew larger it might need support to keep the back or ceiling from collapsing. Supply men (timber men) would lower huge tree trunks and timbers into the mine for support. Some say that entire forests in the Keweenaw were moved underground.

Splitting Copper Blocks - Two miners swing hammers by candlelight under the supervision of a white garbed Quincy mine captain in 1890. *(QMC/HAER,MICH,31-HANC,1-123)*

As the ore was blasted free, gravity would help move it down to the bottom of the stope. Here workers, called trammers, would load it into tram cars used to move the ore to the shaft. Tramming was one of the least skilled jobs in the mine and one of the lowest paying, often one dollar a day. From 1856 until 1862 trammers used wheelbarrows to move the ore to the shaft. In 1862 small 1 ton cars were laid on narrow gauge tracks in the drifts. However, the trammers had to push the cars to the shaft by hand.

Once at the shaft the ore would be hoisted to the surface. In the beginning, rock would be loaded in large metal buckets called kibbles. Kibbles were first raised by hand winches, but Quincy soon began to lift, or drag them with a powered hoist. This allowed the carrying capacity of the kibble to increase from 1,000 to 2,000 pounds of rock. The empty weight of the kibble also rose from 600 to 1,400 pounds. Chains were used to hoist the kibbles until 1864 when Quincy began to use wire rope. In 1864 kibbles gave way to the skips that ran on tracks similar to those seen in the No. 2 shaft-rockhouse today.

Technology and Mining

If a new technology might increase profit then Quincy explored it. When the first mechanical drills arrived in the Keweenaw in 1868, Quincy tried them. It took several trials until a successful one was found in 1879. These large, 600 pound, machines ran on air compressed by a steam driven compressor. It took two miners to move and operate the drill. One of the important tasks was to keep a stream of water at the collar of the drill hole in order to reduce the dust created by the drill. Still there were some miners doing hand drilling in the Quincy until 1903. The mining company decided to use a smaller, 160 pound drill that automatically injected water through a hole in the center of the drill steel. This not only stopped dust but also flushed cuttings from the hole. The new drill was not only more efficient, but now one miner could do the work of two. This was a major cause of the famous 1913 strike in the copper country. After a long and bitter strike, the one man drill became

Scoop Loader - A scoop loader similar to the one on display at work in one of the Calumet and Hecla mines, circa 1960s (*MTU/CCHC, C&H Collection, Neg. 02328*)

Blacksmith Shop (left) and **Machine Shop** (right) - These buildings, shown circa 1900's, are being restored and joined for a new museum. *(MTU/CCHC, Nils Eileetson, donor, Neg. 03387)*

widely used in the copper mines. Several interesting drills, such as the "stoper" or "widow maker," may be seen on the underground mine tour.

Change occurred in many areas of the mine. In 1896 Quincy began to replace candles with "sunshine" or "moonshine" lamps that burned a paraffin based fuel. In 1912 they began to replace these with carbide lamps, and later electric lamps. Quincy never illuminated the mine with electricity. When the lode flattened to 34 degrees as it went lower, gravity alone would not remove the ore from the stope. Mechanical scrapers and tuggers helped move ore from the stope into the tram cars. Tramming became mechanized with the use of electrically powered locomotives in 1901 and later with cars that automatically dumped into pockets (bins, chutes). These pockets were cut into the rock above the shaft to hold the ore for faster loading of the skips. Skip gates held the skip in place while being filled. Blasting practices improved with new explosives, such as dynamite and timed fuses. As early as 1875 Quincy began exploration via horizontal diamond drilling. Quincy was a leader in adopting new technologies.

Quincy's peak employment was 2,000 workers in the period from 1909-1911. By 1945 it was only 425. The mine closed September 1, 1945 due to the high cost of production and the drop in the price for copper as the government ended the subsidy that was in place during the war. With the closing came the end of an amazing chapter in Michigan's mining history.

Looking South from No.2 in the 1920s: From the left, the No.4 boiler house with the Q&TLRR roundhouse behind it, the No. 7 shaft, US 41, the 1860 Blacksmith shop, the 1860 Dry House. The insert shows a similar shot taken in August, 2000 showing the remains of the No. 4 boiler house, the No. 4 hoist and the No. 7 boiler house. *(MTU/CCHC, Nils Eilertson, donor,Neg. 03368, Insert-LJM)*

The Extended Site Tour

Tour Note: This is a continuation of the site tour from page 10 of the tour book. Be sure to examine the intermediate pages as you approach the Nordberg hoist.

As you walk south along US 41 the first sign you'll find will mark the location of the **Man-Engine Shaft** (Man Shaft). Quincy was the third mine in the Keweenaw to have a man-engine, following the Pewabic and the Cliff. From 1856 until this shaft was built in 1866, miners had to climb ladders to enter and leave the mine. This shaft was used until 1892. Although the shaft is covered by grates one may still peer into it and imagine what it must have been like to ride the man-engine, an alternating set of rising and lowering steps, into the darkness each day to go to work. While jumping from platform to platform in the dim light of candles mudded on your hat might seem risky, it was a great improvement over the 725 feet of ladders it replaced. In 1895 man cars (cars with seats) replaced the man-engine and the shaft was left open for ventilation.

As you continue your walk, you'll note a row of signs along US 41. These note the location of the various shafts of the Quincy Mining Company. If you drive the length of the Quincy location you'll note there are two No. 6 shafts. The one you're passing is the original Quincy No. 6. The No. 6 at the north end of the location is the Pewabic No. 6 that Quincy acquired when it purchased the lands of the Pewabic Mining Company.

On your walk you will pass the remains of several Quincy structures. The No. 4 Shaft, operated from 1856 - 1909, is now a concrete pad lying south of the Man-Engine shaft. Behind it, and a bit north lies, the remains of 1882 **No. 4 Boiler House** for the shaft, and behind and a bit further back the remains of 1885 **No. 4 Hoist House** for the shaft. South of this hoist house are the remains of the 1898 **No. 7 Boiler House** for Shaft No. 7. If you walk a straight line from the railroad car loading area of No. 2 to the No. 7 shaft you'll be on the general line of the old **Quincy Tram Road** roadbed. The photograph on page 36 shows the dual tracks running to the shafts and rockhouses. When the Quincy, as well as the Pewabic and Franklin, had their stamp mills in Hancock, they each used a tram road to get the ore to the mill. The weight of the loaded car going down the hill pulled a cable that raised an empty car up the hill back to the mine.

No. 7 Shaft –Rockhouse - Dated about 1905, this photo, taken from the lower end of Quincy Hill looking back towards No. 2, shows the No. 7 hoist house behind No. 7. *(HAER MICH,31-HANC,1-41)*

The Q&TLRR Water Tower as it stands today. See the Site Tour on page 1 for it's exact location and another Q&TLRR photo. *(LJM)*

A short walk south of the No. 7 boiler house leads to the **Q&TLRR Engine House** for the Quincy and Torch Lake Railroad (Q&TLRR). The railroad was created by the Quincy Mining Company to connect its mines with its stamp mill on Torch Lake, where it had an abundant supply of water for milling and a place to dump 'waste' sands. There is a short history of the Q&TL, including many photographs, beginning on page 48 of this tour book. The roundhouse originally contained two engine stalls in 1890. An additional stall was added in 1894. A fourth stall and a machine shop were added in 1900. A rather unique wooden covered **Q&TL water tower**, in the shape of a shaft house, is also on the site. (See page 45 for an original photograph of the water tower.) You may have to walk around a bit to find the tower as it is often hidden by the trees that are attempting to reclaim the site.

If you walked from the engine house area back towards Quincy Hill and crossed No. 7 Flats road, on the line of the **Quincy Tram Road** grade you'd eventually go down the edge of Quincy Hill into East Hancock. All that remains of the tram road is a slight depression in the hill, overgrown by brush and trees. During the 1860s this tram railroad ran from the Quincy mine to its stamp mill on Portage Lake, in the general vicinity of where the Ramada Inn in Hancock is today. The grade comes out just west of Dunstan Street near Lakeview, or west of Reservation Street. The tram ran right through residential areas. Accidents and fatalities occurred when adults, and children, walked the

Agent's House - Here is a winter's view of the house built in 1881*(LGK/HAER,MICH,31-HANC,1-227)*

Agent's House Today - A current view of the house, without the snow. *(LJM)*

tracks and were hit by tram cars.

At the intersection of No. 7 Flats Road and US 41 you'll see a series of buildings across US 41 that were once all part of the Quincy Mine complex. At this time all of them are privately owned. One of the most famous is the red sandstone building, the **Quincy Office Building (Pay Office)**. Begun in 1895 and built of local Jacobsville sandstone, the building cost $29,247. Just inside the door is a large plaque dated June 15, 1917 that honors William R. Todd for his 58 years of service to the Quincy Mining Company, and for serving as its President from 1902 to 1924.

As you look down Quincy Hill from the office building you'll see a home surrounded by trees. This is the **Quincy Agent's Residence**. The agent, or Superintendent was in complete control of the Quincy complex. Completed in 1881 at a cost of over $25,000, this was the most expensive company home built by the Quincy Mining Company. With towering ceilings and magnificent woodwork it was approximately 25 times as costly as a miner's home.

QMC Office Building - Completed in 1897, this photo shows the building in 1911. *(MTU, Roy Drier Collection/HAER,MICH,31-HANC,1-219)*

A Gathering of Engines - In 1908 the Q&TL engines gathered near the engine house. From left to right are #4, #1, #2, #5, and #3.*(MTU/CCHC, Q.M.H.A.)*

Walk past the engine house area and head back along the old Quincy and Torch Lake Railroad grade to the 1920 and 1895 hoist houses directly in a line behind the No. 2 shaft. The Quincy Mine Hoist Association now uses these buildings for their tour of the Nordberg Hoist, the world's largest steam hoist. South of the house is the **No. 5 Boiler House**. The boilers here supplied steam to No. 2 Hoist. It replaced Boiler House Number Four, about 1,000' from No. 2 Hoist House, and was connected by a 14" steam line. The building was built in 1912 for a cost of $19,400. It housed eight boilers and has a 144-foot high smokestack. In front of the hoist house, and between it and the No. 2 shaft-rockhouse, is the No. 2 Outside Display Area.

Your tour will continue by walking the drive from the hoist house to Lower Pewabic Road. As you walk you will pass a house typical of many of the miner's homes. It is under reconstruction by the Q.M.H.A.

The few homes to your far right are all that remains of the community of **Lower Pewabic**. Many mining companies offered low rent housing as an incentive for their employees. The Pewabic Mining Company began building Lower Pewabic in the 1860s. Quincy acquired the land and added additional homes. Several of these homes were built according to plans supplied by

The Water Tower - Q&TL at the water tower near the engine house in 1910-1915.*(LGK/HAER,MICH,31-HANC,1-101)*

Sears, Roebuck, and Company in 1918. Recent research has shown that company supplied housing did not achieve its desired goal and that most employees left it for other housing within a short time. Most of the homes in Lower Pewabic have been demolished. There are a few of the Sears and Roebuck homes left in the western half of the subdivision.

The location map shows many of the small communities that grew up around the mine. While the names may sound ethnic, their population pattern was not. That is, Limerick was not just for the Irish, and you did not have to be French to live in Frenchtown. Most of the homes in these locations have now been demolished.

Walk towards the junction of Lower Pewabic Road and US 41. While fields border the road now, the photograph of the water cooling ponds on page 25 shows the homes that once bordered the street. As you near US 41 you will see the **Machine Shop** on your right. It was built in 1900. The brick building rests

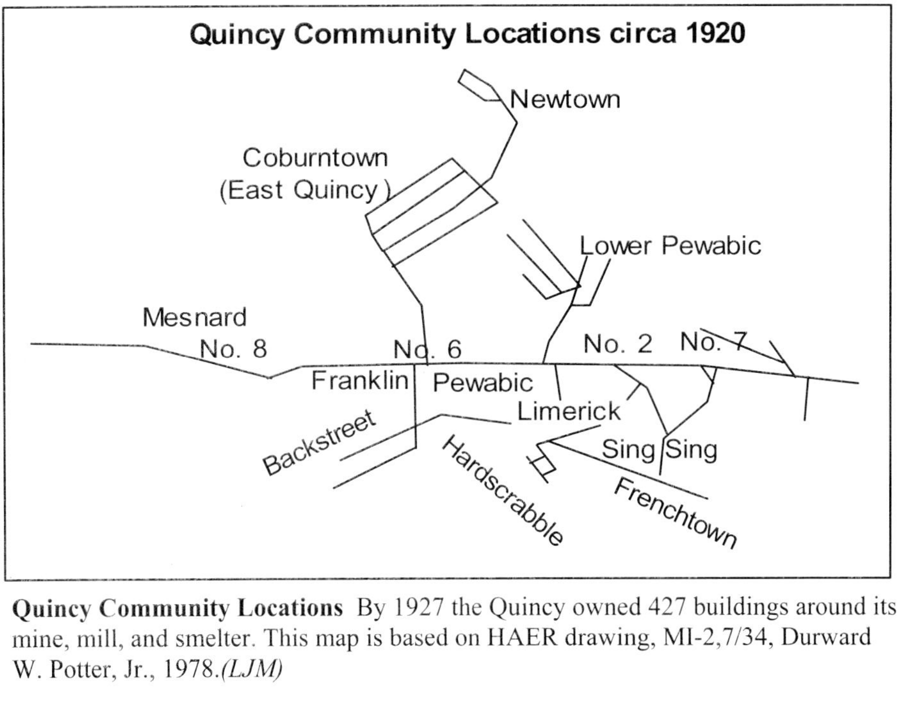

Quincy Community Locations By 1927 the Quincy owned 427 buildings around its mine, mill, and smelter. This map is based on HAER drawing, MI-2,7/34, Durward W. Potter, Jr., 1978.*(LJM)*

Looking North from No.2 in the 1920s: From the left is the Limerick location across US 41 and the 1893 supply office (lower left, now the Q.M.H.A. gift shop). Behind it is the 1893 carpenter shop and a lumber shed. Further back is the 1900 blacksmith shop, the 1900 machine shop, and the No. 6 Compressor building. In the far back is the1892 No. 6 Shaft-Rockhouse and the No. 6 and No. 2 Boiler Houses. Note the many railroad tracks that crossed the area. The insert shows a similar shot taken in August, 2000 showing the supply office, machine shop, blacksmith and the stacks from the boiler houses. *(QMC/HAER,MICH,31-HANC,1-86, Insert-LJM)*

on a red sandstone foundation. Just north of the machine shop is the **Blacksmith Shop,** also built in 1900. It originally contained twelve forges, steam hammers, grindstones, etc. When built it was called "the model smithy of the copper district." Both of these buildings are undergoing a rebirth. Michigan Technological University is restoring and renovating both buildings to be the future home of the A. E. Seaman Mineral Museum, the Mineral Museum of Michigan. When the Machine Shop adaptive restoration is complete, now in progress, one of North America's finest mineral collections will be on Quincy Hill next to the famous Quincy mine which produced great mineral specimens.

Further north, along US 41, lies the remains of the "North Quincy," or **No. 6 Shaft**, originally part of the Pewabic Mining Company. Surrounded by a border of poor rock and covered by wire mesh stretched over a grating of rails lies the collar of the Number Six Shaft. The shaft was operated from 1891 - 1931, and then from 1937 - 1945. The shaft opening has recently been modified with the addition of a bat cage. This interesting arrangement of metal allows bats to enter and exit their underground quarters and yet, keeps visitors from entering the shaft. Behind No. 6 are the ruins of **No. 6 Dry House**. Built before 1895 and remodeled in 1907, this dry, or change house for miners to change into and out of their work clothes, served the number six shaft. Behind the dry, in the woods, stand the smokestacks of the No. 6 and No. 2 boiler houses that provided steam to this end of the complex. A walk through the area will also show some slightly elevated rail lines, part of the wide array of railroad tracks that crisscrossed the mine location.

Further north from No. 6, beyond the usual walking distance, stands a small, metal hoist house and an open air, metal skeleton of a head frame. The **Quincy-Homestake Project** (Todd Project) explored the **No. 8 shaft** here. The shaft can be clearly seen from US 41. The land was originally purchased from the **Mesnard** Mining Company by Quincy in 1897 and operated to 1931. It then operated intermittently from 1937 to 1945. In 1976 Quincy and the Homestake Mining Company began an exploration project on Quincy Mining Company land. The shaft was cleared to the 20th level. Diamond drilling for exploration purposes continued at the site through 1977. No production was attempted. When the project stopped in 1977 it marked the end of exploration or mining for the Quincy Mining Company.

Quincy No. 8 Shaft - The man-car and headframe now stand idle after the Homestake project closed in 1977. (*LJM*)

Turn south and begin to walk back to the No. 2 shaft-rock house. On the southeast corner of Lower Pewabic Road and US 41, opposite the fire station, stands a weathered building. It was the **Mine Captain's Office**. In the early era the mine captain was in charge of all operations. He was often an experienced miner who not only could 'read the rock' to follow the path of the copper ore, but also managed mine operations. Later, as mine captains became college-educated geologists and engineers, the mine captain became more focused on underground operations. The location of the office, in the center of the mine site, provided easy access to all areas of the mine.

As you continue to No. 2 you'll walk past the community of **Limerick** across US 41 from the mine location. Your walk will pass the locations of several old mine buildings, including the carpenter's shop which once stood just north of the supply shop (Q.M.H.A. gift shop). The gift shop marks the end of the extended site tour.

The Quincy and Torch Lake Railroad

To rail fans the most interesting items in the Hoist House Outdoor Display Area are two locomotives of the Quincy and Torch Lake Railroad. The Q&TLRR was built not only from rails and spikes, but also by corporate pride, questionable trials, and carefully crafted definitions. The Q&TLRR might be looked at as an extension of the original Quincy tram road. The tram used cars balanced in pairs so that as the car full of ore went down Quincy Hill to the mill on Portage Lake empty cars were pulled up the hill. This

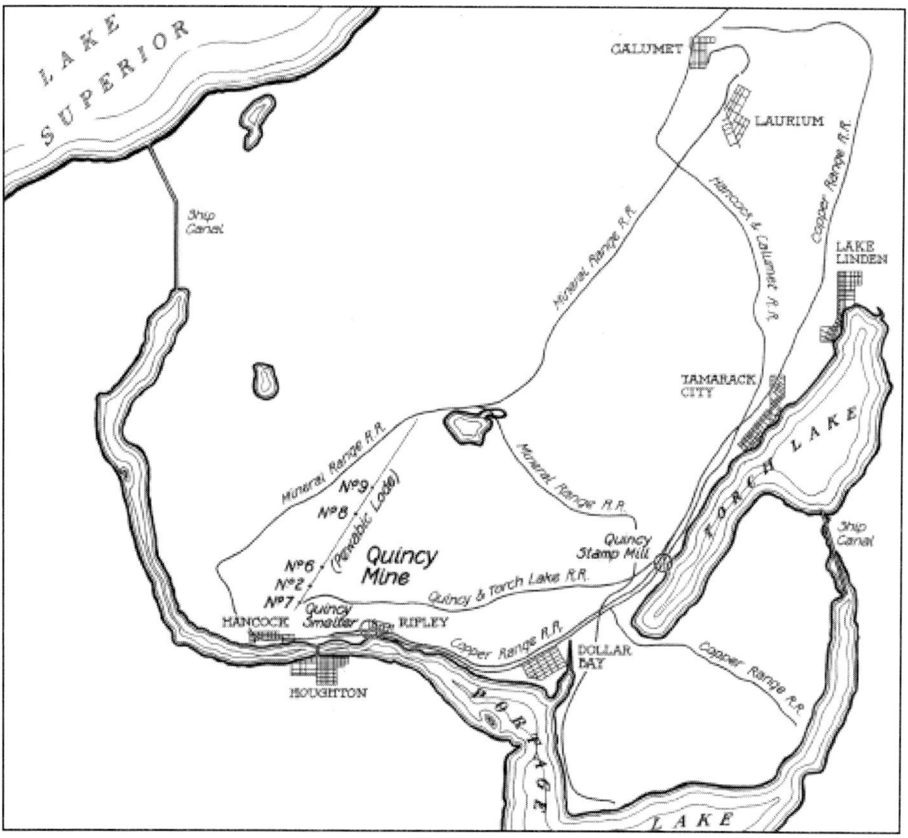

Quincy Mining Company Locality - Based on a map in the 1910 QMC report, this map shows the route of the Q&TL and other railroads. This map is part of HAER drawing, MI-2,2/34, Durward W. Potter, Jr. , 1978 and Richard K. Anderson, Jr. , 1979. *(LJM)*

Q&TL #1, The Thomas Mason, built by Brooks in May 1889, as shown here, as #1535, a 2-6-0, class 15-B-3, 32 ton Mogul, 15 x 20" cylinders, 42" drivers, now on display, with the tender from #2, at the Q.M.H.A. steam hoist. It was the first Q&TL to receive dual gauge coupler pockets and only the one with power operated pilot truck flanges. *(MTU/CCHC, Q.M.H.A.)*

gravity powered system served the mine from the late 1850s to 1890. However, by 1887 it had become apparent to both the Quincy Mining Company and the U.S. government that the stamp sand (tailings or waste) from the mill that was being dumped into the lake was beginning to hinder navigation. Quincy decided to build a new mill approximately 6 miles away on the shores of Torch Lake, where they had miles of shoreline that allowed them to dump the sand into the lake. But how would the ore get from the mine to the mill?

Quincy was already contracting rail service to haul coal to the mine, etc., from two local railroads. Both the Mineral Range Railroad (MRRR) and the Hancock and Calumet Railroad (H&CRR) were already serving Quincy, although on a cautious basis. Quincy already had problems with the MRRR in 1872 when it denied the MRRR right-of-way through its Portage Lake property. Quincy began negotiations with both railroads to haul their ore to the new mill. But as early as January, 1888, Quincy began to plan its' own railroad. Part of this effort was a bargaining ploy for negotiations, but it was also spurred by a corporate desire for a railroad. S.B. Harris, the Agent for the Quincy, called the possibility of a Quincy run railroad a 'glorious thing' in April, 1888. The negotiations began to fail and the Q&TLRR was organized under the laws of Michigan on June 27, 1888. By July 5 the negotiations had failed and the decision was made to begin construction of the Q&TLRR.

Q&TL #2, built by Brooks in November 1889, as #1577, a 2-6-0, class 15-B-3, 32 ton Mogul, 15 x 20" cylinders, 42" drivers. It went out of service around 1910. The engine was scrapped in 1915 at which time the tender went to engine #5. The tender is now on display behind engine #1 at the Q.M.H.A. steam hoist. *(MTU/CCHC, John F. Campbell Collection, Q.M.H.A.)*

One issue that stood in the way of the railroad was fact that Quincy did not own the land between the mine and the mill. As a "public" railroad they negotiated right-of-way from several landowners. However, the owners of the Pewabic and the Franklin mines, lying just north of the Quincy, refused. The controversy over the request for a right-of-way was just one more item on the list of issues that would lock these competitors into legal combat for years. Eventually the Quincy took control of the Pewabic in 1891 and the Franklin in 1908. While the Q&TL was organized as a public railroad, there were never any plans to carry anything other than Quincy freight. In fact, 494 shares of it's total stock of 500 shares was owned by Thomas F. Mason, President of the Quincy.

When the Pewabic and Franklin refused to negotiate a right of way, the Quincy took them to court. Quincy asked for public condemnation proceedings to gain the right-of-way it needed. The case bounced from the local court, to the Michigan Supreme Court, and back to the local court. A jury that included a brother of the man who had already been selected to build the railroad finally decided the case in Quincy's favor. Quincy now could begin the railroad. However, the argument that they were a public railroad almost came back to haunt Quincy in 1925. Quincy then appealed a tax assessment on the Q&TL by stating that since it had never moved any freight other than that owned by the Quincy, the railroad could not possibly be a public railroad and could not be taxed as such. While this argument seems to be 180 degrees from the earlier court case, the court accepted the argument and the assessment was reduced.

One of the major decisions that faced the builders of the Q&TL was that of the gauge or space between the rails. While rail gauge is standard in America today it was not in 1888. Should the Q&TL be standard gauge, with 4' 8 1/2" between the rails? The Duluth, South Shore and Atlantic was a standard gauge railroad and had discussed serving the Quincy. Or should the Q&TL be a narrow gauge railroad with 3' between them? Both the H&CRR and the MRRR were narrow gauge. The decision was made to use narrow gauge in order to connect to these two lines. Clearing the right-of-way began in August 1888 and track laying in April 1889, using 60 pound rail on 8-foot ties. The only structures built for the Q&TL were the round house and a water tank at the mine site (both of which still stand) and another water tank at the mill, some six miles away. Two turntables, one at the mine and one at the mill were also installed. The rolling stock began with 30 rock cars, three flat cars, and one caboose in August 1889. The Thomas F. Mason, Q&TL#1 arrived in June 1889. It's near twin, Q&TL#2, arrived in November 1889. While #1 ran until

Q&TL #3, Shown here as it was built by Brooks in October, 1894, as #2475, a 2-6-0, class 17-B-3, 45 ton Mogul, 17 x 22" cylinders, 44" drivers, out of service at end of operations in 1945. In 1979 it was leased to the Huckleberry Railroad, Flint, Michigan. *(MTU/CCHC, Ray W. Buhrmaster Collection, Q/M.H.A.)*

Q&TL #4, Shown here as it was built by Brooks in October, 1900, as #3639, a 2-6-0, class 17-B-3, 45 ton Mogul, 17 x 22" cylinders, 44" drivers, out of service by the 1940's and scrapped at the end of operations. *(MTU/CCHC, Ray W. Buhrmaster Collection, Q/M.H.A.)*

1945, #2 had many mechanical problems and was scrapped in 1915. Regular operations began on March 20, 1890. Approximately 42 cars per day went from the mine to the mill.

After the acquisition of the Pewabic and Franklin properties the Q&TL expanded north so that any building that needed to receive coal or ship ore was serviced by the Q&TLRR. A 6.5 mile rail line grew to over 14 miles just by adding sidings and spurs. The array of rails, switches, and yards formed a spider web across the Quincy property. When Q&TL#3 arrived in 1894 an additional stall was added to the roundhouse. Q&TL#4 arrived in 1900. However #4 was plagued by problems that Quincy blamed on the Brooks manufacturing company. Consequently, Q&TL#5 came from a different manufacturer. In fact, #5 was a 10 year old Baldwin engine that the Q&TLRR purchased from the H&CRR. When it roamed the Quincy property for the H&CRR it was known as the "Opechee". Now it became Q&TL#5. This engine was involved in the only major accident on the line. On November 15, 1937 a trestle north of Dollar Bay collapsed due to rain-soaked ground. The engineer and fireman were both killed. Q&TL#5 was rebuilt and ran until the end of service in 1945. In 1913 the last and largest of the Q&TL engines arrived. While #3 and #4 could haul 24 empty rock cars from the mill to the mine, #6 could handle up to 44 cars.

Unlike the tram, the Q&TLRR was designed to move freight both ways. The old coal yard had been built in 1888. The H&CRR and MRRR had carried the coal from the old coal yard on Portage Lake to the mine. However in 1902 a new coal unloading facility was built near the new stamp mill on Torch Lake. It's three coal unloading towers could unload 13,500 pounds of coal per minute. The two railroads were not able to handle the increased demands of the Quincy. Quincy would have to take over shipments both up and down Quincy Hill. In 1902 a new rail spur was run from behind the mill down to the new coal shed located on the shores of Torch Lake. Other changes to the Q&TLRR came from technological advances. For example, by 1909 almost all of the Q&TL's 140 pieces of rolling stock were equipped with the newly announced automatic air brakes. This improvement not only increased safety, but it also increased production as each car could now carry more weight as it now assisted the engine in stopping the train.

The Q&TLRR did not carry all of the freight for the Quincy company. It ran only from the mine to the mill. In 1898 Quincy built its smelter in Ripley, near Hancock. How would the material get from the mill some five miles

Q&TL #5, The "Opechee", built by Baldwin in January, 1891, as #11534, a 2-8-0, Consolidation, 16 x 20" cylinders, 37" drivers. It was originally built for the Hancock & Calumet R.R. as their #7, renumbered to #34 in 1893, and purchased by the Q.&T.L. in February, 1908 as their #5. Now on display at the Q.M.H.A. steam hoist with the tender from #1.*(MTU/CCHC, John F. Campbell Collection, Q.M.H.A.)*

Q&TL #6 shown circa 1935-37. Built by Baldwin in December, 1912, as #38920, a 2-8-0, Outside framed Consolidation, 18 x 20" cylinders, 36" drivers. Leased, in 1975 to the Pine Creek R.R, Allair, N.J.*(MTU/CCHC, John F. Campbell Collection, Q.M.H.A.)*

back along the shoreline to the smelter? The DSS&A, which had taken control of the H&CRR and the MRRR, finally agreed to build a standard gauge line connecting the mill and smelter if the Quincy would buy standard gauge cars. Thus Quincy was in an unusual position of supporting both standard and narrow gauge equipment.

By 1905 the Q&TLRR was running about six twenty-car trains per day in each direction, with the odd numbered trains running from the mine to the mill and the even numbered the reverse route. In order to keep track of expenses that would be billed back to the Quincy, several forms were completed for each train, and there would have been at least 15 forms completed each day. One of the few surface improvements on the line occurred in 1912 when the two turntables were replaced by "Y's" that allowed the engine to switch direction without the maintenance involved with the turntables, especially in the winter snows.

Since 1894, the Q&TLRR had charged the Quincy a rate for transport that equaled it's monthly operating costs. However, the railroad still had large debts for construction, equipment, etc. On April 4, 1905 Quincy purchased all of the equipment from the Q&TLRR and left it with only the ownership of the right-of-way. Quincy would lease the track for a monthly fee. From that point on the Q&TLRR no longer was a railroad and was just another part of the mine. After 1920 it is no longer mentioned in the Quincy Annual Reports. In 1927, following the tax case of 1925, it was decided that the Q&TLRR track and right-of-ways should be sold to the Quincy Mining Company. On February 1, 1927 the railroad sold it's last assets and ceased to be a legal entity. The history of the company ends at that point, but the railroad continued. Trains ran up to the time the mine closed in 1931 and then reopened with the mine in 1937. On September 1, 1945 the mine closed for good and the Q&TLRR engines were placed in their roundhouse for the last time.

The Q&TLRR is gone but it is not forgotten. During the summer of 1975 #1 and #5 were moved to their present location near the hoist while #3 and #6 were moved to other railroad museums.

Q&TL#5 - This photo, taken in 1942, shows #5 running past the coal-car trestle at the #5 boiler house that served the No.2 hoist. *(MTU/CCHC, John F. Campbell Collection, Q.M.H.A.)*

References

The Condensed History of the Quincy & Torch Lake Railroad, Kevin E. Musser, http://www.portup.com/copperrange/qtl.hthttp://www.portup.com/copperrange/qtl.htm

Copper Mines of the Keweenaw, Don H. Clark, Wood Duck Art , No. 15, Quincy Mining Company (1980)

Cradle to Grave, Life, Work, and Death at the Lake Superior Copper Mines, Larry Lankton, Oxford University Press, 1991

Dead But Not Buried, Franklin A. King, Trains Magazine, Jan 1960

The Hard Rock Mining Era In The Copper Country, Tauno Kilpela, 1995

Historic American Engineering Record, Report of Quincy Mining Company, HAER MI-2, 1978, including:

An Economic and Business History of the Quincy Mining Company, Charles K. Hyde:

Technological Change at the Quincy Mine c. 1841-1931, Larry D. Lankton

Quincy Mining Company: Housing and Community Services, c.1860-1931, Sarah McNear

Quincy Mining Company: Stamp Mills and Milling Technologies, c.1860-1931, Charles F. O'Connell

A History of the Quincy and Torch Lake Rail Road Company, 1888-1927, Charles F. O'Connell

Keweenaw Copper: Geology and History, Theodore J. Bornhorst & Larry D. Lankton, Great Lakes Geoscience LLC, 2006.

Library of Congress, Prints and Photographs Division, Historic American Engineering Record, http://memory.loc.gov/ammem/hhhtml/hhhome.html

Locomotives of the Quincy & Torch Lake, John F. Campbell, Narrow Gauge & Short Line Gazette, May, 76

Michigan's Historic Sites Online, Michigan Historical Center, State Historic Preservation Office, http://www.state.mi.us/mdos/michsite

Old Reliable, an Illustrated History of the Quincy Mining Company, Larry D. Lankton & Charles K. Hyde, Quincy Mine Hoist Association, 1982

The Quincy & Torch Lake - A railroad built for modeling, Kevin E. Musser, http://detnews.com/metro/hobbies/hotbox/spring98/quincy/quincy.htm

Quincy & Torch Lake Moguls 1 & 4, Ed Gebhardt and Paul Meier , Railroad Model Craftsman Aug 1986

Quincy Mining Company Housing 1840's-1920's, Nancy B. Fisher, unpublished Masters Thesis, Michigan Technological University, 1997

Quincy Mining Company, Mine, Smelter, Community, Quincy Mine Hoist Association

Range of Opportunity, A Historic Study of the Copper Range Company, Richard A. Fields, Quincy Mine Hoist Association, 1997

Self-guided Geological Field Trip to the Keweenaw Peninsula, Michigan, T. J. Bornhorst and W. I. Rose, Institute on Lake Superior Geology Proceedings, Volume 40, Part 2, 1994

Self Guided Walking Tour Quincy Mine National Historic Landmark, H. L. Spencer and T. S. Reynolds, Program in Industrial History and Archeology, Dept. of Social Sciences, MTU, SIA Conference, 1997

The QMHA Tram Skillings Mining Review, August 23, 1997, Vol. 86, No. 34

The Upper Peninsula of Michigan, An Inventory of Historic Engineering and Industrial Sites, Historic American Engineering Record (HAER), Charles K. Hyde, 1978

Valve Gear Program, version Q.1, written by Charlie Dockstader It is available from several Internet sites.

Wonderful Power, Susan R. Martin, Wayne State University Press, 1999

Index

My Tour Notes:

The date I took the tour: ______________________________

The weather was ____________________________________

Things I'll remember about the :
The No. 2 Shaft– Rock house Tour

The Quincy Mine surface Tour

The Nordberg Hoist House Tour

The Q&TLRR Tram

The Underground Tour

Other things I found interesting: